Vegetable Forcing

Containing Information on Greenhouse Construction, Management and Frame Culture

By

Ralph L. Watts

British Library Cataloguing-in-Publication Data
A catalogue record for this book is available from the
British Library

Farming

Agriculture, also called farming or husbandry, is the cultivation of animals, plants, or fungi for fibre, biofuel, drugs and other products used to sustain and enhance human life. Agriculture was the key development in the rise of sedentary human civilization, whereby farming of domesticated species created food surpluses that nurtured the development of civilization. It is hence, of extraordinary importance for the development of society, as we know it today. The word *agriculture* is a late Middle English adaptation of Latin *agricultūra*, from *ager*, 'field', and *cultūra*, 'cultivation' or 'growing'. The history of agriculture dates back thousands of years, and its development has been driven and defined by vastly different climates, cultures, and technologies. However all farming generally relies on techniques to expand and maintain the lands that are suitable for raising domesticated species. For plants, this usually requires some form of irrigation, although there are methods of dryland farming. Livestock are raised in a combination of grassland-based and landless systems, in an industry that covers almost one-third of the world's ice- and water-free area.

Agricultural practices such as irrigation, crop rotation, fertilizers, pesticides and the domestication of livestock were developed long ago, but have made great progress in the past century. The history of agriculture has played a major role in human history, as agricultural progress has been a crucial factor in worldwide socio-economic change. Division of labour in agricultural societies made (now) commonplace specializations, rarely seen in hunter-gatherer cultures, which allowed the growth of towns and cities, and the complex societies we call civilizations. When farmers became capable of producing food beyond the needs of their own families, others in their society were freed to devote themselves to projects other than food acquisition. Historians and anthropologists have long argued that the development of agriculture made civilization possible.

In the developed world, industrial agriculture based on large-scale monoculture has become the dominant system of modern farming, although there is growing support for sustainable agriculture, including permaculture and organic agriculture. Until the Industrial Revolution, the vast majority of the human population laboured in agriculture. Pre-industrial agriculture was typically for self-sustenance, in which farmers raised most of their crops for their own consumption, instead of cash crops for trade. A remarkable shift in agricultural practices has occurred over the past two centuries however, in response to new technologies, and the development of world markets. This also has led to technological improvements in agricultural techniques, such as the Haber-Bosch method for synthesizing ammonium nitrate which made the traditional practice of recycling nutrients with crop rotation and animal manure less important.

Modern agronomy, plant breeding, agrochemicals such as pesticides and fertilizers, and technological improvements have sharply increased yields from cultivation, but at the same time have caused widespread ecological damage and negative human health effects. Selective breeding and modern practices in animal husbandry have similarly increased the output of meat, but have raised concerns about animal welfare and the health effects of the antibiotics, growth hormones, and other chemicals commonly used in industrial meat production. Genetically Modified Organisms are an increasing component of agriculture today, although they are banned in several countries. Another controversial issue is 'water management'; an increasingly global issue fostering debate. Significant degradation of land and water resources, including the depletion of aquifers, has been observed in recent decades, and the effects of global warming on agriculture and of agriculture on global warming are still not fully understood.

The agricultural world of today is at a cross roads. Over one third of the worlds workers are employed in agriculture, second only to the services sector, but its future is uncertain. A constantly growing world population is necessitating more and more land being utilised for growth of food stuffs, but also the burgeoning mechanised methods of food cultivation and harvesting means that many farming jobs are becoming redundant. Quite how the sector will respond to these challenges remains to be seen.

VEGETABLE FORCING

VEGETABLE forcing is the production of vegetables in greenhouses, hotbeds, coldframes, or other structures. In frame culture in the spring or fall, glass may be used during only a part of the period of growth.

The business of vegetable forcing is highly specialized. The purpose of this discussion is to present briefly the fundamental principles and practices.

Competition is severe and is augmented by the improved methods of packing and transportation which supply fresh vegetables from warmer regions throughout the winter season. Under the conditions few growers consider it profitable to expand their glass beyond the area needed for plant growing.

Forcing boxes are the simplest means of advancing the growth of crops. In effect they are miniature coldframes which are placed over certain crops, for example hills of melons or cucumbers, immediately after sowing seeds or setting plants. They are especially useful in regions where the transition period from cool to warm weather is long, and where the summers are too short to produce satisfactory crops of warm-season vegetables. They are used to a limited extent also to advance the harvest in warmer regions. A very few growers use forcing boxes extensively. They usually are made as rectangular, bottomless, light wooden boxes to be covered with a single pane of 10x12-inch glass which slides in grooves to permit ventilation. The use of forcing boxes is described on page 454.

Plant protectors.—In a sense, the various more or less transparent plant protectors, in the form of cones or domes of paper or similar material, also may be considered as small forcing structures. Their use is similar to that of forcing boxes, although they are much less expensive and not so laborious to apply. They are not generally used in the East, but a few growers who have become skilled in their management find them profitable with extra early crops, especially cucurbits or tomatoes. As the days become warmer, ventilation is provided by slitting the tops.

Frame culture is practiced by many market gardeners, and is extensively carried on around certain shipping points within the belt of mild winter weather from Norfolk, Virginia, southward along

J. W. Davis Co.

FIG. 143.—One of the largest vegetable forcing ranges, heated from a central power plant. Houses in the different groups are 425 to 700 feet long.

the seaboard. Both cloth-covered and sash-covered frames are used. Lettuce is the most popular frame crop and radishes are widely grown, sometimes as an intercrop between hills of cucumbers or plants of the solanaceous crops. Beets, cucumbers, muskmelons, tomatoes, and cauliflower often are planted in frames several weeks before they could be started in the open.

FIG. 144—Removing hotcaps from early lettuce with three plants set under each cover. Caps are used most extensively in starting melons and, to a considerable extent, tomatoes and other tender crops.

Newly made hotbeds, spent hotbeds, or coldframes may be used, depending upon the time of planting, the crop, and available facilities. Parsley and lettuce are favorite frame crops for late fall. Little capital is required to engage in frame culture with a few sash, and the enterprise can be expanded as profits justify. Liberal manuring and fertilization, and thorough control of insects and diseases are essential. Frame culture of cucurbits is discussed on page 457.

GREENHOUSE CONSTRUCTION

Greenhouses are in general use among market gardeners. The first house, which perhaps is very small, is built for the purpose of starting early vegetable plants, for which it is found convenient and satisfactory. The owner, however, is often unwilling to have it idle more than half the year; therefore, he tries a forcing crop. If his efforts in the production of crops under glass prove successful, the greenhouse area is increased and new houses are built from year to year, until the grower is known as a vegetable forcer. Greenhouses furnish better conditions for starting early plants and they may be used 10 or 11 months in the year if the establishment is properly handled. It is not uncommon for market gardeners to operate an acre or more of greenhouse space, while a much larger proportion of growers have from 1,000 to 10,000 square feet of glass.

If forced vegetables can be grown profitably, which is more difficult than formerly when shipping methods and facilities were poorly developed, greenhouses will provide several advantages. With winter crops the grower keeps in touch with the market the whole year round. The houses are ideal for plant growing, and provide pleasant employment in the winter when it often is difficult to find sufficient work to keep men busy.

The size.—The proper size of a greenhouse is determined by a number of factors. It is never safe to build extensively without thorough experience in greenhouse work and marketing. A house 30x100 feet is probably as large as any market gardener should start with, and a smaller structure would be desirable where both capital and experience are limited. A width of 30 feet has been stated, because this is about the minimum width for economical construction, heating, and operation; narrower houses do not provide as uniform atmospheric conditions, and the plants are more likely to be injured by direct cold drafts.

Location and position.—While the natural protection of buildings, trees, or hills on the north and the west sides is highly desirable, greenhouses should not be constructed where they will be shaded.

The position of the house with reference to the points of the compass is not of major importance. With the modern house, ample light will enter with any orientation; most growers, however, have the length of the house running northeast and southwest when it is possible.

Types of construction.—Serviceability, durability, and economical construction and operation are the main points to keep in mind when building greenhouses. Full iron frame construction may or may not be the most economical in the end; the first cost is from one-third to one-half greater than for semi-iron construction, and this additional expense may exceed the cost of repairs in other types. With proper care and painting the wood parts in a well-built house will last 25 years.

Semi-iron type of construction.—This is by far the most popular form of construction. The walls are usually concrete below the side glass and the pipe or angle iron posts which support the roof are placed on concrete footings. If provision is made for walks or alleys along the walls in the greenhouse the side posts should extend at least 6 feet above the ground level. All purlins, braces, and interior posts are made of iron. Plates, ridges, and rafters are of durable wood, usually cypress. However, angle iron eaves are especially desirable. The construction is of such character that it is easily possible to replace decayed wood parts without disturbing the posts, purlins, or braces. Walks, beds, and benches of concrete may be used if desired. Houses of this type are attractive, serviceable, and durable. Any good carpenter can build them without difficulty, for nearly all wood parts are cut to the right dimensions at the factory and blue prints are furnished for the instruction of builders. Manufacturers of materials supply helpful catalogues and handbooks.

Forms of greenhouses.—Lean-to, even-span, and three-quarter-span are the principal forms of greenhouse construction. The lean-to house is the least expensive type to construct, and it is most useful when erected against a building or wall running northeast and southwest, where there will be practically no shading of the plants. As it is simple in construction, it is popular among home gardeners and small-scale growers.

Even-span houses are the most satisfactory for commercial growers. Sash houses (p. 36) are used to a limited extent in forcing vegetables.

Walls.—Whatever material is used in their construction, the walls should be started below frost line. Concrete walls banked with earth on the outside are warm and inexpensive to construct, and are much better than those built of wooden posts and boards. Although cedar and locust posts last for many years, the lower boards in a wall will soon decay. If the foundation extends about 2 feet above grade level, there will be a height of about 4 feet of glass at the sides between the plate on the concrete wall and the gutter or eaves plate, where sufficient height is desired for a walk next to the wall.

Roof construction.—The size of the roof bars is determined by the width of the glass, and the distances between posts, braces, and purlins and should be short enough that sagging does not occur. An approved method of bracing and standard pitch should be used. In the ridge and furrow plan of construction, gutters must be provided between the houses; these, however, deteriorate quickly unless they are made of rust-resistant metal. Many growers prefer to build with a space of 10 to 15 feet between the houses. This not only renders gutters unnecessary, but reduces shading in the houses and allows room into which snow may slide from the roofs.

Glass.—Experience has taught greenhouse men that anything inferior to A grade double-strength glass should not be used. The lighter and inferior grades not only sustain greater breakage from hailstones, but the imperfections may cause injury to the plants by burning.

Small glass increases the cost of construction and decreases the amount of light. Glass measuring 16x24 is the standard size today, and is used to a much greater extent than any other size. It is generally laid with the sash bars 16 inches apart. If the panes are graded before glazing so that lights of about equal curvature are placed together, there will be very little space between the laps. A distance of 20 inches between sash bars is regarded as proper by some growers, in which case 20x24-inch glass is used.

Glazing and painting.—A priming coat of paint should be applied to all wood parts before construction is started. The glass then is bedded in a glazing compound or in putty consisting of one part of white lead to five parts of putty. After puttying the shoulders of the sash bars the glass is placed with the curve up, and pressed down firmly, squeezing out the surplus material. This method of glazing is standard because the putty remains in place and keeps out water and cold air and prevents the escape of heat from the house. Glazing points should be inserted at the laps. After the glass has been laid the house should receive two additional coats of paint; it should be repainted every year to insure maximum durability. Later it may be advisable to "bulb" the house with a glazing compound placed in the outside angle between glass and sash bars.

Ventilators.—Provision must be made for ample ventilation. The most approved plan is to have a line of vents on both sides of the ridge. When houses are used until midsummer or later, side ventilators often are provided. It is of the greatest importance that the ventilating machinery work properly.

Beds, benches, and walks.—At one time vegetable forcers thought it essential to provide benches with bottom heat for practically all greenhouse crops, but the opinions held today are different. In many of the largest and most successful vegetable forcing ranges there are no benches or even ground beds with board, brick, or concrete sides. The houses often have large doors at the ends so that a horse and cart can enter with manure or other supplies. It is also possible to use plow and harrow in the preparation of the soil for planting.

Benches are convenient for handling flats and potted plants, but they are expensive to construct and maintain unless made of concrete and they waste space. In some cases they may be needed to supply bottom heat for starting warmth-loving plants.

Solid beds, with or without sides, provide more uniform moisture conditions than do raised benches and there is less danger of injury from improper watering. Solid beds are especially advantageous when the watering must be intrusted to men of limited experience.

The walks should be arranged so that all the beds can be cared for conveniently. Beds or benches 5 to 8 feet wide and walks or alleys 18 inches in width make a desirable combination, although the width of beds in large commercial houses is generally much greater. As plants do not thrive next to the walls, it is desirable to have walks there, and the house space can then be divided in such manner as may seem convenient for the care and harvesting of the crops to be grown.

Steam versus hotwater heating.—Hot water is unquestionably the best system of heating small houses because the pipes retain heat for a greater length of time. Thus the boiler may be left for longer periods without attention, a matter of great advantage in small greenhouses, where it would not pay to employ a night fireman.

On the other hand the steam heating system requires less radiation and is, therefore, less expensive to install. Steam also may be piped with less difficulty to a distance for auxilliary heating of hotbeds or cold frames and is useful for disinfecting soil (p. 44).

The boiler.—The boiler should be of ample capacity to maintain proper temperatures; forcing a boiler means waste of fuel, and the boiler itself will not last long. The construction should be of such type that the greatest heat will be realized from the fuel consumed.

Radiation.—Most frequently 1½ and 2-inch pipe is used for the coils in hot water heating and 1¼-inch pipe for steam heating, connecting with mains of proper size. Whatever the system, the pipe should be placed with the greatest care, observing the principles of the method of heating to be used. Inexperienced builders can secure detailed instructions and plans from the firms supplying construction materials. Farmers' Bulletin 1218 of the United States Department of Agriculture is useful in planning the system.

The work room.—In every greenhouse establishment there is a great deal of work to be done in the way of seed sowing, transplanting, potting, and preparing crops for market. A convenient, commodious work-room is a necessity. The room should be well lighted and properly heated and ventilated. Tables of the right height are necessary, and a small room for tools will be found very convenient.

GREENHOUSE MANAGEMENT

Because conditions in the greenhouse are artificial, the success of the crops depends upon skill and regularity of management. Suggestions in Chapter IV under the topics: Soil supply (p. 43), Soil disinfection (p. 44), Watering (p. 56), and Ventilation (p. 57) apply equally in general greenhouse management.

Soils.—Naturally deep, well drained, friable soils are desirable, but the original character of the soil in greenhouses is changed so rapidly that the exact type is not particularly important. Both heavy and light soils, as originally found, are common in greenhouses. Because several crops, often lettuce, tomatoes, and cucum-

The E. Bigelow Co.

FIG. 145.—Four-inch drain tile header in place, ready to connect 3-inch drain tile circulating lines joined by tees at the distant ends.

bers, may be grown in the same house in the course of a year or two, the soil should be of generally suitable type.

In small houses it may be most convenient and economical to replace the soil every year or two.

In commercial ranges the soil usually is disinfected as often as necessary and retained indefinitely. The buried tile system of steaming the soil, as illustrated, is employed most extensively. Details of installation and operation can be secured from the Agricultural Experiment Stations of important vegetable forcing states, of which Ohio is first, as well as from certain manufacturers of tile, and well informed growers. Two especially important points are admission of steam to both ends of the header and connection of the distant ends of the lateral lines to assure even distribution and circulation. Disinfection of soil with formaldehyde and small scale use of steam have been considered (p. 44).

Manures and Fertilizers.—Because of the very high overhead costs in vegetable forcing, both manure and fertilizer should be used in optimum amounts. Maintenance of excellent physical condition and light texture requires yearly applications of 25 to 50 tons of manure to the acre—about one ton to a thousand square feet. The usual practice is to apply all the manure before the fall crop and to work it thoroughly and deeply into the soil.

At one time manure was used almost exclusively in the fertilization of crops grown under glass, but it is now recognized that many troubles of greenhouse crops are due to inadequate, often unbalanced, nutrition. The finer points, particularly timely top-dressing of cucumbers and tomatoes with nitrogen or potash, are learned only by study, observation, and experience. Actual practices vary considerably with crops, soils, and methods of the individual grower (pp. 491, 495).

As a general rule, 1000 pounds to the acre, that is 25 pounds to 1000 square feet, of a suitable fertilizer can be applied with advantage before each crop and serve to supplement the application of manure as suggested. Fertilizer applications should be worked into the soil thoroughly and evenly. Concerning analyses, 5-10-5, or 4-12-4 if very much manure has been applied, are often used before lettuce, cucumbers, and fall tomatoes. When tomatoes are planted during the dark months, 0-10-10 or the equivalent is recommended. Top-dressing is discussed in connection with tomatoes and cucumbers (pp. 491, 495).

Lime.—Although it is unwise to permit high acidity to develop in the greenhouse soil, excessive liming should be avoided. Some growers apply 1000 pounds of dolomitic limestone to the acre yearly, a practice which is not likely to lead to difficulties either way. Actual testing for lime requirement is advisable.

Watering.—The principles described for plant growing are fundamental guides, but it should be emphasized that greenhouse vegetables suffer more frequently than is realized from lack of water. Dry subsoil at the time of summer or fall planting is a serious handicap. Long continued watering, preferably before planting, may be required. Again, the fruitful period of spring tomatoes and particularly of cucumbers often is curtailed, and yields are reduced because of inadequate supplies of water. Excessive watering also is very detrimental, especially in late fall and winter.

Temperatures.—Definite temperatures cannot be established as ideal. Slightly higher temperatures may be desirable to establish newly set plants quickly and somewhat lower temperatures than normal may be desirable under some conditions to retard over-vegetative growth. The temperatures indicated below may be considered normal for average conditions. The range given embraces the ideas of different growers and the usually permissible variations.

NORMAL TEMPERATURES IN VEGETABLE FORCING
(In degrees Fahrenheit)

Crop	*Night*	*Cloudy Days*	*Sunny Days*
Lettuce	45-50 (55)	50-55	55-60 (70)
Tomatoes	60-65	65-70	70-80
Cucumbers	65-70	70-75	75-90
Radishes	43-48	50-55	55-65
Rhubarb	50-60	(Usually grown in darkness)	
Mushrooms	50-60	(Higher temperatures are objectionable)	

Day temperatures occasionally 10 to 15 degrees higher than those suggested are not likely to prove detrimental on sunny days with free ventilation and low humidity. Long continued temperatures below normal may cause stunting from which the crop, especially cucumbers, may not fully recover. Too high temperatures are followed by soft, spindly, succulent growth, and in extreme cases by dropping of blossoms and should be avoided if possible.

Mulching with strawy manure is common practice with tomatoes and cucumbers but may reduce yields at seasons when much nitrogen is needed by the growing crop. This objection may be overcome by heavier feeding with nitrogen (p. 75).

Fumigation is the most convenient and effective means of combating most insects that cause injury to greenhouse crops. The first step in fumigating is to calculate the volume of the house. The cross sectional area (the area of an end) in square feet is multiplied by the length in feet; the product is the number of cubic feet of space in the house which should be determined with accuracy. Broken glass must be replaced and all other openings closed. No applications of water should be made within 24 hours as the plants must be perfectly dry and the humidity low. It is preferable that the plants grow "on the dry side" for a few days. An even or slightly rising temperature during fumigation will avoid condensation. Temperatures of 60 to 70° F. are suggested. Fumigation may begin an hour or more after sundown. Observance of these precautions is necessary to avoid burning the more tender portions of sensitive plants. To avoid wastage of gas select a calm night, but not a rainy one.

Hydrocyanic acid gas is effective against aphids, white flies, and thrips, but it must not be used after applications of bordeaux mixture or copper lime dust. These fungicides may be applied after hydrocyanic fumigation.

The simplest method of fumigating with hydrocyanic acid gas is to scatter calcium cyanide thinly and evenly on the walks. Liberation of the deadly gas begins immediately. The correct dosage varies from 1/8 to 1/4 ounce to 1,000 cubic feet, depending upon conditions. It is safer to start with the smaller dosage and increase if necessary at the next fumigation. Fumigation every 10 to 14 days will secure effective control.

Because hydrocyanic acid gas is extremely poisonous it should never be used to fumigate greenhouses that are connected with buildings in which human beings or domestic animals are present during the treatment. Avoid breathing the least trace of either the dust or the fumes. Make certain all people and pets are out of the house before spreading the cyanide, and that the exit is open. If there is more than one walk one person should be assigned to each. All workers should begin at the same time at the end farthest from the door, proceed quickly at the same rate, and leave the house together. Doors should then be closed and locked, and signs posted at all entrances reading "Danger—Fumigating with Poison Gas." Never go back over a fumigated area immediately, or reenter the house until the next day. Then open the doors and ventilators to air the house.

Empty houses sometimes are fumigated for a period of 24 hours with one or two pounds of calcium cyanide to 1,000 cubic feet to "burn out" pests that may be present. Temperatures of 70° F. or higher are recommended.

Fumigation with nicotine which is effective against aphids is convenient and safe but usually more expensive than hydrocyanic acid gas. Nicotine usually is employed, however, where aphids are the principal pest. It may be applied in a number of simple ways, and manufacturers will supply full information on dosages and procedure.

Fumigation by burning sulphur after the last of the crop has been harvested will kill exposed insects and spores of fungus diseases. Sublimed or refined sulphur burns more readily and is recommended at the rate of 1 pound to 2,000 cubic feet for a 24-hour fumigation. Greater dosages are unnecessary and undesirable, on account of their effect on paint, pipe, and wires.

A sulphur vaporizing machine advertised as the Sulfur Nebulator provides a means of mild fumigation that is useful in controlling leaf mold of tomatoes when the treatments are begun in the early stages.

PRINCIPAL FORCING CROPS

Lettuce.—The Grand Rapids variety of leaf lettuce is planted almost exclusively in greenhouses. Butter heading varieties, such as White-seeded Tennisball, May King, and Boston Market or Belmont, are grown to a limited extent, particularly around Boston. Cos or Romaine lettuce is produced for special demand. New York, commercially known as Iceberg, lettuce has not proved profitable on account of the long period of growth and usually excellent condition and low price of the western shipped product.

March Automatic Irrigation Co.

FIG. 146.—Irrigating greenhouse lettuce.

Fertilizers and temperatures (pp. 486, 487) for greenhouse lettuce have been suggested. Spray irrigation is permissible for this crop.

Lettuce plants must be ready to set in the beds at the desired planting dates or valuable time and space will be lost. Seed for the earliest practical fall crop usually is sown in shaded houses or outdoors about the middle of August. The seedlings are transplanted about 2 by 2 inches apart, and are shifted to the beds as soon as they are of desired size. Later sowings will be required to provide plants for resetting the house, and it should be remembered that six weeks may be required to grow plants of suitable size during the winter months. Popular planting distances are 7 by 7, 7½ by 7½, 8 by 8, 7 by 8, and 7 by 9 inches.

Harvesting leaf lettuce in less than six weeks from setting in the beds is unlikely to be as profitable as when more time is given. After a certain point, increases in weight take place very rapidly. Over-maturity reduces table and market quality, however, and may result in losses to the grower. Harvesting early may be advantageous to make way for a succeeding crop or to sell retail by count. Lettuce diseases were discussed (pp. 275, 276).

Radish.—This crop demands maximum light and will not do well in dark houses at midwinter, or as an intercrop where it is subject to shading. The quickly-growing, short-topped selections of Scarlet Globe usually are grown. High temperatures (p. 487) or crowding will result in excessive development of leaves. Spacing 4 to 5 inches between rows, and sowing or thinning to secure 8 to 10 plants to the foot of row, are standard. Four to 8 weeks, depending upon the amount of sunshine and length of days, are required to mature the crop.

Tomato.—Because this crop requires uniform warmth (p. 487) and is subject to leaf disease its culture during the fall and winter should be undertaken only in first-class houses that are well-lighted, not leaky, and easily heated. The spring crop is less difficult to manage. The long period of growth requires effective control of insects, especially the white fly.

Varieties of the Bonny Best class, including John Baer and Chalk Jewel, are generally popular. Globe is desirable where pink fruits are wanted. Marvel, Break O'Day, Marglobe and Marhio are useful where resistance to wilt is essential. Varieties of English ancestry, such as Comet, Carter Sunrise, and Grand Rapids forcing, produce many-fruited clusters of comparatively small tomatoes and are preferred where the markets readily accept small fruits.

The fall crop yields most heavily when started in time for a number of clusters to be set before dull, winter weather begins. The seed usually is sown late in June or early in July and well grown plants are set in the greenhouse not later than the middle of August.

The midwinter crop is the most difficult to grow and produces disappointing yields without the best management and equipment. Seed for it is sown in August, September, or October, depending on when the plants are to be set in the beds.

Plants for the spring crop usually are set during February which requires the sowing of seed about December 1. The methods of plant growing should be such that sizeable, vigorous plants are produced. Low-grade plants lengthen the unfruitful period. Plants transplanted once into 4-inch pots or bands are ideal (pp. 285, 453).

Popular planting distances are 18 by 36 to 20 by 42 inches. Many growers prefer spacings of 16 to 18 inches, however, in rows about 3½ feet apart, on account of greater convenience in working among the plants and better circulation of air between the rows. Crowding increases the expense of plants, because greater numbers are required, provides more favorable conditions for disease, and is more likely to decrease than to increase the total yield.

As a guiding principle in fertilizing greenhouse tomatoes, and to a certain extent cucumbers, nitrogen should be provided in liberal supply in the spring and during the long sunny days of summer, but it is required in relatively limited amounts during the short, usually dark days of late fall and winter. It is commonly thought that excess nitrogen results in failure to set fruit during prolonged cloudy weather and, under the same conditions, liberal fertilization with potash promotes normal ripening and aids the development of solid, well-filled fruits (p. 100).

Fall tomatoes set in the beds during August may receive about 1000 pounds to the acre, or 25 pounds to 1000 square feet, of 5-10-5 before planting. Two or three weeks later about 5 pounds of nitrate of soda to 1,000 square feet, or the equivalent, can be applied as top-dressing. Additional nitrogenous top-dressings at intervals of about two weeks may or may not be beneficial, but it is customary to make the last one some time before the period of dark weather begins. By that time the plants should be well set with fruits. Thereafter occasional top-dressings with potash, as 4 or 5 pounds of muriate of potash to 1,000 square feet, are in order, at least on some soils.

The spring crop of tomatoes, if it is planted during the dark weather of winter, needs little or no nitrogen in the pre-planting application. Then the analysis 0-10-10 may be used, however, a fertilizer of 1-2-1 or 1-2-2 ratio would be preferable in some instances. When days become longer and sunny, nitrogenous top-dressings may be instituted and are necessary during the bearing season of spring crops to secure highest yields.

The present trend is to fertilize more heavily than formerly. This has resulted in much higher yields and notably better quality of fruit in the later pickings. A synopsis of recent recommendations of the Ohio Agriculture Experiment Station follows:

(1). Apply about 50 tons of manure to the acre in advance of the fall crop.

(2). In preparing the soil for planting (spring crop tomatoes) work in plow-deep 1000 to 1500 pounds of superphosphate and 750 to 1000 pounds of muriate of potash to the acre.

(3). When the first three clusters of blossoms have set and the earliest fruits are half grown, make the first topdressing. Subsequent applications should follow at intervals of 10 to 14 days during the life of the crop. The rate of each is about 250 pounds to the acre.

First application is potassium nitrate.
Then two of calcium nitrate.
One of potassium nitrate.
Two or three of calcium nitrate.
One of potassium nitrate.

Or 100 pounds of potassium nitrate and 200 pounds of calcium nitrate may be mixed and used at the recommendated rate in making each application.

The topdressing material is broadcast between the rows and watered very thoroughly to dissolve all the fertilizer and carry it deeply into the soil. Thus root injury is avoided.

Nearly all experienced growers train tomato plants to a single stem. Beginning as soon as the plants are set, the laterals are snapped or pinched out at frequent intervals.

One of the most common means of support is twine tied around the base of the plants and stretched vertically to wires or rafters overhead. See Figure 147. Thin stakes are preferred by some in which case the plants are tied to the supports at several places as growth proceeds. Leaf pruning is not generally advisable.

Natural pollination of greenhouse tomatoes often is inadequate to produce satisfactory yields. Artificial aid to pollination is practically a necessity during the dark days of winter, often doubles the yield, and produces smoother fruits. Recommendations from Bulletin 470 of the Cornell Agricultural Experiment Station, Ithaca, N. Y., follow:

"It is recommended that greenhouse growers of American tomato varieties pollinate the flowers on the first three or four clusters by the watch-glass method, in late winter and early spring. Flowers on clusters above the third or the fourth should be pollinated by daily jarring. The emasculation method is just as effective as the watch-glass method, but it is more costly and is less practical with American varieties; with English varieties, however, it is recommended for the lower clusters in late winter."

"In the watch-glass method of pollination, a quantity of pollen is collected on a watch glass or a glass slide, from flowers which have expanded petals. The watch glass is then held in the left hand, being placed just beneath the flower with well-reflexed petals, and with the right hand the stigma of the flower is gently brought into contact with the pollen. The pollen might be collected on the thumb nail or on the tips of the fingers of the left hand, as explained under the emasculation method, but the watch glass is probably more desirable. If pollen is applied prematurely, it may lose its vitality before the stigma becomes receptive, and fail to germinate. However, the work of Hartley (1902) indicates that premature pollination does not injure the pistil, so that if the pistil is again pollinated, when receptive, with fresh pollen it will probably be fertilized and fruit will develop. An important advantage of the watch-glass method under commercial conditions is that the supply of pollen on the glass constantly increases as the work progresses, because the pollen may be procured from blossoms to be pollinated by tapping them before pollinating."

"In pollinating plants by emasculation, the pollen is collected on the thumb nail or on the tips of the first and second fingers of the left hand, from flowers with petals well reflexed. Only flowers which have been fully open and have closed their petals are pollinated, after the petals and stamens have been removed. As the stamens are attached to the corolla, both may be easily removed in one operation by grasping the tip of the wilted corolla with the thumb and the first or second finger of the left hand, holding the flower back of the calyx with the right hand. Then pollen is applied to the stigma by lightly touching the stigmatic surface with either the finger or the thumb nail covered with the pollen. The thumb and the first finger of the right hand are used to steady the blossom during the operation. With this method one must be careful, in removing the corolla and the stamens, to pull straight away from the flower, otherwise the pistil may be broken at the base of the style. There is also danger of break-

Department of Vegetable Crops, Cornell University

FIG. 147.—Greenhouse tomatoes, showing supports.

ing the pistil when the finger covered with pollen is brought into contact with the stigma, because the natural support furnished the pistil by the stamens and the corolla has been removed."

"Jarring the plants is the method which practically all commercial greenhouse men use, chiefly because it is the simplest, easiest, and quickest method. The plant or clusters may be jarred in various ways, but the method employed in these experiments was to grasp the main stem of each individual plant and

Department of Vegetable Crops, Cornell University

FIG. 148.—Tomato fruit cluster showing various stages of flower and fruit development.

a, Petals well reflexed; proper stage for pollination by the watch-glass and brush methods.
b, Flowers have been fully open, and petals have closed and begun to shirvel; proper stage for pollination by the emasculation method.
c, Emasculated flower ready for pollination.
d, Fruit developing from flower pollinated by the emasculation method, showing pistil attached but petals removed.
e, Fruit developing from flower pollinated by the watch-glass method, showing petals attached.
f, Unpollinated flower failing to develop a fruit.

shake it sharply. Jarring was always done toward noon, preferably on sunny days when the temperature was high and the atmosphere dry."

To assure high quality, greenhouse tomatoes should not be harvested until they are well colored. Packages of small size are desirable to prevent bruising and crushing. Paper wrappers and special cartons are used to some extent to protect the fruits and to make attractive packages.

Ten pounds to the plant with spring crops and half this amount, or less, with fall or winter crops are typical yields.

Disease control must not be neglected (pp. 294, 486).

Cucumber.—The first requirement for profitable forcing of cucumbers during cold weather is a house in which uniformly high temperatures can be maintained without prohibitive expense for fuel. Control of insects, nematodes (p. 181), and red spiders (p. 179) must be effective. Red spiders often ruin a crop before their presence is observed.

Popular forcing varieties of cucumbers include Davis Perfect, Abundance, Irondequoit, Deltus, and well-bred selections of White Spine under a variety of names. Many growers develop their own selections and crosses (p. 466).

Strong, vigorous plants that have been grown without checking or stunting are required to produce an early, heavy crop. Plants for the spring crop usually are set in the beds during February, although some growers, especially with poorly heated houses, produce an additional crop of lettuce and defer the bedding of the cucumber crop until March or early April.

Seed should be sown only five or six weeks before the plants are to be ready. Clay pots of veneer bands in 3- to 4-inch sizes generally are used. Several seeds may be sown in each container and the plants thinned to one or two; or the seeds may be sown quite thickly in flats of sand, peat, or light compost. In that case the seedlings must be transplanted to the pots very promptly, usually within a week of sowing (pp. 445, 453).

For the growing of cucumber plants and for the crop, normal night temperatures are 65° to 70° and day temperatures 75° to 90°. The day temperatures may be several degrees lower in cloudy weather and, in early summer, may rise to 100° without harm when ventilation and moisture are not neglected.

As a rule, cucumbers respond favorably to liberal feeding. In addition to the usual application of manure, 1,000 pounds or more to the acre, or 25 pounds to 1,000 square feet, of 5-10-5 or the equivalent may be worked into the soil before planting. Top-dressing after the fruits begin to set is nearly always desirable at intervals of about two weeks, or sooner if the plants begin to turn yellow. When plants of the spring crop are bearing very heavily, some growers top-dress every week or 10 days. A mixture of 4 pounds of nitrate of soda and 1 pound of muriate of potash to 1,000 square feet, or an equivalent amount of other fertilizers, may be more effective than nitrogen alone. Suggestions for manuring and fertilizing tomatoes (pp. 491, 492) are quite applicable.

With the A-trellis the plants usually are spaced 12 to 15 inches apart in the rows, with the pairs of rows for a trellis about 8 feet apart, more or less, so that the trellises suit the width of the house. Sufficient space for a narrow walk between trellises ordinarily is allowed, but not always. In the arbor system the plants are trained on twine or light wooden strips until they reach a flat trellis 6 or 7 feet overhead, where they are permitted to spread. Planting distances 3 by 5, 4 by 4, and 4 by 5 are common. The sides of the A-trellis and the horizontal overhead trellis usually are constructed of No. 16 wires about six inches apart.

In either system the plants are pruned to a single stem and tied at intervals until they reach the top of the support, where the main terminal is pinched out. In the A-system usually the laterals also are nipped just beyond the first female blossoms. In the arbor system a suitable amount of pruning is necessary to prevent too dense vining. Although perfect pruning rarely is accomplished, neglect of the operation is not justifiable as long as the market is satisfactory.

Pollination of the cucumber out-of-doors is accomplished by insects that carry pollen from the male to the female flowers. With a few plants, in the amateur's greenhouse, a camel's hair brush may be used to transfer pollen from the male flowers to the female flowers, the ones with miniature cucumbers in the centers.

Bees are used in commercial houses, ordinarily one or two hives for houses 200 feet in length. In cold weather the hive must be kept inside and should be shaded. When warm weather arrives it is customary to place the hives just outside the houses and remove adjacent panes of glass for the bees to enter. Because cucumber blossoms produce much pollen but little nectar, supplementary feeding of the bees is desirable and the colonies should be replaced before they become exhausted.

Cucumbers in the greenhouse should be picked three times a week to secure the fruits at the best marketable stage. Careful grading usually increases returns.

Rhubarb is the most easily grown of the forcing crops. It is produced in cellars of dwellings, beneath greenhouse benches, and in special forcing houses constructed of any convenient material. Strong root-clumps, one-year old or older, are lifted with as little injury as possible to the large storage roots. After lifting, the roots must be subjected to thorough but not extremely severe freezing.

The total rest period should be at least six or eight weeks. The roots may be stored in any shed or may be sheltered outdoors to protect them from drying winds, or they may be placed in position in the forcing house if they can be permitted to freeze and rest there before the heat is turned on.

The clumps are ranked close together when being placed. Light soil or coal ashes is worked into all the spaces. The roots should rest on a bed of soil and be covered to a depth of 2 or 3 inches. Harvesting will be difficult if the beds are more than 4 or 5 feet wide. Very thorough watering is the next step, and additional water should be applied as required to maintain moisture. Temperatures of 50° to 60° F. are ideal.

Other vegetables are forced in greenhouses to a very limited extent. Although many may be grown successfully, their culture generally is unprofitable. Forcing of Witloof chicory or French endive is discussed on page 278.

Mushroom culture.—Successful production of any crop depends upon correct management of many details. This is especially true in the culture of mushrooms. Extensive production should not be attempted without experience. Authoritative literature is available and should be studied by those who are especially interested. The cultural directions which follow are only elementary.

Cool cellars, the space beneath benches in cool-crop greenhouses, unused buildings, caves, or mines, can be adapted without much expense to small-scale growing of mushrooms. The commercial crop is produced almost entirely in specially constructed houses. The most important characteristic of a suitable place for growing mushrooms is relatively constant temperature averaging about 55° F. For the first few weeks after the beds have been planted with spawn the temperature may be maintained at about 70° F. This allows rapid development of the mycelium without injury. After the casing soil has been added the temperature should be reduced to 60° F. or lower. Temperatures below 55° F. are not injurious but result in slower growth. After the mushrooms have begun to appear, temperatures above 60° F. should be avoided as they allow the development of diseases and insect pests. At 70° F. the mushrooms on the beds begin to show injury and will soon die if the temperature remains at or above that point for more than a few hours. While the most important characteristic of a suitable place for mushroom growing is the proper temperature, an adequate method of ventilation cannot be neglected. This is necessary to prevent the humidity from becoming too high and to avoid the accumulation of carbon dioxide which is given off abundantly by the growing mushrooms and the decaying manure.

Manure for mushroom beds almost invariably is prepared by

processing fresh horse manure from animals that have been fed a diet containing grain and bedded with straw. A considerable proportion of straw is desirable.

About 3 or 4 weeks before the beds are to be filled the manure is placed in rectangular, flat-topped piles with nearly vertical sides. At intervals of about 5 days the pile is turned, and rebuilt as described for hotbeds (p. 78). At each forking-over, all lumps must be shaken apart, the outside of the old pile placed inside the new one, and water added to dry spots. After 3 or 4 turnings the compost should be uniformly dark brown in color, friable, and free from unpleasant odors. The straw should be easy to break when a bunch is wrung between the hands. It should not, however, have lost its heat. In this condition it is ready to be placed in the beds. The moisture content at this stage is important. A test for this is to squeeze a handful of the composted manure. It is too wet if drops of water can be squeezed out, but the hand should become distinctly wet. If the pile is exposed to the weather it is advisable to protect it from heavy rains toward the end of the composting period.

Prolonged composting is very objectionable as the material becomes too fine and loses its resilience, resulting in a lumpy condition after it is placed in the beds. Also the organic matter is being continually lost from the pile as carbon dioxide, thus removing nutrients valuable to the mushrooms.

In this country flat beds in tiers are almost universally used, while in Europe ridged beds are placed directly on the floor. The former are much superior as they are not in contact with the floor—the lowest bed is built 6 inches above the floor—and it is possible to provide much more area to the unit of building volume. The manure is placed 5 to 7 inches deep in the beds. The present practice is not to tamp the compost very much, although some growers still press it firmly into the beds. The temperature should rise at first. Commercial growers try to obtain a temperature of at least 120° F. in the air of their houses and 130° F. in the beds. This is sufficient to kill the insects and disease organisms which otherwise would menace the crop later on. If the type of building is such that this heating is impossible after the beds are made up, the pile of compost should be large enough so that the cold exterior of the pile harboring most of the insects can be discarded and only the hot part used. After filling the beds, if the compost is too dry, several very light waterings may be applied.

Spawning must be delayed until the temperature subsides to about 70° F. as determined by a thermometer plunged in the beds. Further delay is disadvantageous. The best spawn can be procured directly from growers or specialists, who are in the spawn-production business, and should be fresh. Spawn more than a few weeks old, as sometimes sold by dealers, may be worthless.

Three kinds of spawn are available. These are: the older type made from manure in bottles, that using tobacco stems as a base, and finally that made from grain. A satisfactory crop may be grown with any of the three. The manure spawn can be removed only by breaking the bottle. It is then broken into pieces about the size of walnuts and buried two inches deep in the beds at 10-inch intervals. The tobacco stem spawn is crumbled to fine particles, while the grain spawn is already granular. These are planted at intervals in the beds as is the manure spawn. The grain spawn is more vigorous than the others, one bottle being sufficient to plant 100 square feet of bed. In about 10 to 14 days, but no sooner, the beds should be cased with one inch of rich, moist, medium loam soil, the kind that is ideal for plant growing.

Heavy watering is harmful. The rule with mushrooms is to water lightly and frequently to keep the casing soil moist but not wet and to avoid water penetrating to the manure beneath. Humidity should be high enough to prevent rapid drying of the casing soil, but not excessively high.

Light is not injurious, except for the effect of direct sunlight in raising temperatures.

Production usually begins in 5 to 10 weeks after spawning and continues for 3 to 6 months. Warm weather, where temperatures can not be held down, will shorten the productive period.

Daily picking is necessary to secure the mushrooms as they attain the desired size, before the veil joining the stem and the edges of the cap has broken. Mushrooms usually are picked by twisting the base of the stem. The holes that are made in picking should be closed with casing soil.

Mushrooms should be marketed promptly after picking to avoid unnecessary deterioration. Great care must be used to avoid bruising and excess handling. Packaging is important for protection of the product. In the Middle West pint and quart paper boxes with cellophane tops are finding especial favor, although some pound boxes are used. In the East 1 and 4-quart climax baskets are accepted as standard, being packed to hold 1 and 3 pounds of mushrooms. For the local market quart berry boxes will be found satisfactory.

The following recommendations for the control of mushroom diseases, weed fungi, and insects are by W. S. Beach, Bulletin 351 of the Pennsylvania Agricultural Experiment Station:

"Mushroom houses can be made sanitary by fumigation with formaldehyde, 3 lbs. to 1,000 cu. ft., or burning flowers of sulphur, 2 lbs. to 1,000 cu. ft. To assure an effective concentration of the fumes requires tight sealing of all cracks and ventilators, rapid evolution of the fumes, and thorough diffusion. These requirements are more likely to be attained if the fumes are evolved outside the house with an apparatus designed for the purpose and then blown inside near the floor.

"Fumigation needs to be supplemented by drenching floors and spraying doorways and bottom beds with a fungicidal solution, especially if fumes of formaldehyde or sulphur are evolved within the house. One of the following may be used: 4 lbs. of copper sulphate dissolved in 50 gals. of water, 2½ per cent crude carbolic icid, or 16 lbs. of formaldehyde to 50 gals. of water. Spraying all inside surfaces and structures with a fungicidal solution can be substituted for fumigation. These treatments are also suitable for packing sheds.

"Treat adjacent yards, composting ground, and other exterior parts of the plant that may have become infested with fungous spores, using one of the following fungicidal solutions at the rate of at least a quart to the square foot: 8 lbs. of copper sulphate dissolved in 50 gals. of water, 7 ozs. of mercuric chloride dissolved in 50 gals. of water, or 2½ per cent crude carbolic acid.

"Deposits of spent compost or waste mushroom fragments should not be left near a mushroom house, as they are ready sources of disease.

"Before refilling is begun, empty, clean, and disinfect, as far as possible, all houses that may be infested with disease or weed fungi. This is particularly important in a series of adjoining houses.

"A temperature of 130° to 140° F. after filling is necessary to kill any harmful fungi introduced with the compost. Electric fans set horizontally at the top of the center aisle are usually essential to overcome the differences of temperature between top and bottom beds. Avoid unnecessarily high temperatures, above 140° to 150° F.

"Observe sanitary precautions about houses from the time the compost cools for spawning so that the spores of fungi will not be carried to clean beds. Wash contaminated hands and clothes, and disinfect soiled baskets or other utensils. This is important for the exclusion of the *Verticillium* disease (brown spot), *Mycogone* disease (bubbles), bacterial blotch, and white plaster mold.

"Sterilize the soil for casing, or be certain that it is obtained from a field that has not been contaminated with spent mushroom manure, waste, or drainage from houses having disease. While this applies particularly to the *Mycogone* disease, it is also important in preventing the introduction of other diseases or molds.

"Endeavor to maintain the temperature of bearing beds within the range that is optimum for mushrooms, 55° to 58° F. Many diseases and weed molds become serious on account of abnormally high temperatures. Avoid warm season crops unless an efficient cooling and ventilating system is installed.

"Provide efficient ventilation so that excessive humidity and precipitation of moisture in bearing houses may be readily removed or prevented. A desirable humidity is 88 to 90 per cent. Ventilation is most efficient when the outside temperature is below that inside. Surplus moisture upon mushrooms in the pinhead and button stages should dry readily if brown spot and bacterial blotch are to be held in check.

"Watch for the first occasional or local development of brown spot, bubbles, or mildew. Pick all good mushrooms within an area of disease, and remove and destroy all diseased specimens. Spray the diseased area with a dilute bordeaux mixture (1-1-50 or 2-2-50). This will not injure later flushes and will hold the diseases in check. It is possible to follow this same procedure when brown spot is general. It may prove more effective than recasing with fresh soil. Copper-lime dust, containing 10 or 15 per cent monohydrated copper sulphate and no arsenical, may be substituted for bordeaux mixture.

"Control of readily disseminated weed molds, particularly white plaster mold and olive-green *(Chætomium olivaceum)*, depends chiefly on favorable composting of the manure. If the spawn starts and grows well, it will take possession of the compost in spite of considerable contamination by these molds. If the composting is not favorable and the spawn is retarded, the molds are likely to predominate. It is advisable to test the reaction, or pH, of the compost just before the beds are cooled for spawning to see whether the heating has reduced the pH low enough for the spawn to grow readily. The nearer the pH is to neutral (pH 7.0) the more likely it is that the spawn will outgrow the molds.

"Control of heat-resistant weed molds, like truffle, requires a thorough disinfection of the composting ground and other contaminated places. Spray the interior of infested houses with a fungicidal solution.

"Insects are important carriers of fungous spores, hence control of insects is an aid in checking the spread of harmful fungi.

"Filling houses but once a year is a distinct advantage in the control of harmful fungi because the mushroom crop can be grown entirely within a cool season, when temperature and moisture conditions can be easily kept optimum; many fungi, especially *Verticillium* and white plaster mold, disappear largely by drying during the months that the houses are empty; and there is less likelihood that picking will continue in infested houses while other houses are being refilled."

Solution culture of vegetables, using chemical nutrients in water, is attracting widespread popular attention. Several methods have been devised, of which a typical one is to support the plants on litter-covered screens at the top of shallow tanks containing the solution. Formulae for the solutions vary. As growth takes place composition and reaction of the solution change. Tests must be the basis for correction. It may be necessary to modify the solution to meet the differing needs of the plant as growth progresses or as the length of day changes. The services of a trained plant physiologist are essential for consistently successful production of crops in nutrient solutions. At the present stage of development such methods apparently do not provide commercially valuable advantages over culture in soil.

Sand culture of plants before transplanting into the field is discussed on page 58.

CPSIA information can be obtained
at www.ICGtesting.com
Printed in the USA
LVOW11s0011260917
550043LV00001B/95/P